>>>————————————————<<<

THIS BOOK BELONGS TO

>>>————————————————<<<

A B C D E F G H I J K L M N O P Q R S T U V W X Y Z

WEBSITE

EMAIL USED

USERNAME

PASSWORD

NOTES

WEBSITE

EMAIL USED

USERNAME

PASSWORD

NOTES

WEBSITE

EMAIL USED

USERNAME

PASSWORD

NOTES

WEBSITE

EMAIL USED

USERNAME

PASSWORD

NOTES

WEBSITE

EMAIL USED

USERNAME

PASSWORD

NOTES

WEBSITE

EMAIL USED

USERNAME

PASSWORD

NOTES

A
B
C
D
E
F
G
H
I
J
K
L
M
N
O
P
Q
R
S
T
U
V
W
X
Y
Z

A B C D E F G H I J K L M N O P Q R S T U V W X Y Z

WEBSITE

EMAIL USED

USERNAME

PASSWORD

NOTES

WEBSITE

EMAIL USED

USERNAME

PASSWORD

NOTES

WEBSITE

EMAIL USED

USERNAME

PASSWORD

NOTES

WEBSITE

EMAIL USED

USERNAME

PASSWORD

NOTES

WEBSITE

EMAIL USED

USERNAME

PASSWORD

NOTES

WEBSITE

EMAIL USED

USERNAME

PASSWORD

NOTES

A
B
C
D
E
F
G
H
I
J
K
L
M
N
O
P
Q
R
S
T
U
V
W
X
Y
Z

A
B
C
D
E
F
G
H
I
J
K
L
M
N
O
P
Q
R
S
T
U
V
W
X
Y
Z

WEBSITE

EMAIL USED

USERNAME

PASSWORD

NOTES

WEBSITE

EMAIL USED

USERNAME

PASSWORD

NOTES

WEBSITE

EMAIL USED

USERNAME

PASSWORD

NOTES

WEBSITE

EMAIL USED

USERNAME

PASSWORD

NOTES

WEBSITE

EMAIL USED

USERNAME

PASSWORD

NOTES

WEBSITE

EMAIL USED

USERNAME

PASSWORD

NOTES

A
B
C
D
E
F
G
H
I
J
K
L
M
N
O
P
Q
R
S
T
U
V
W
X
Y
Z

A
B
C | WEBSITE |
D
E EMAIL USED

F USERNAME

G PASSWORD
H

I NOTES

J | WEBSITE |
K
L EMAIL USED

M USERNAME

N PASSWORD
O

P NOTES
Q

R
S | WEBSITE |
T
U EMAIL USED

V USERNAME

W PASSWORD
X

Y NOTES
Z

WEBSITE

EMAIL USED

USERNAME

PASSWORD

NOTES

WEBSITE

EMAIL USED

USERNAME

PASSWORD

NOTES

WEBSITE

EMAIL USED

USERNAME

PASSWORD

NOTES

A
B
C
D
E
F
G
H
I
J
K
L
M
N
O
P
Q
R
S
T
U
V
W
X
Y
Z

A
B
C
D
E
F
G
H
I
J
K
L
M
N
O
P
Q
R
S
T
U
V
W
X
Y
Z

WEBSITE

EMAIL USED

USERNAME

PASSWORD

NOTES

WEBSITE

EMAIL USED

USERNAME

PASSWORD

NOTES

WEBSITE

EMAIL USED

USERNAME

PASSWORD

NOTES

WEBSITE

EMAIL USED

USERNAME

PASSWORD

NOTES

WEBSITE

EMAIL USED

USERNAME

PASSWORD

NOTES

WEBSITE

EMAIL USED

USERNAME

PASSWORD

NOTES

A
B
C
D
E
F
G
H
I
J
K
L
M
N
O
P
Q
R
S
T
U
V
W
X
Y
Z

A
B
C

WEBSITE

EMAIL USED

USERNAME

PASSWORD

NOTES

D
E
F
G
H
I
J
K

WEBSITE

EMAIL USED

USERNAME

PASSWORD

NOTES

L
M
N
O
P
Q
R
S

WEBSITE

EMAIL USED

USERNAME

PASSWORD

NOTES

T
U
V
W
X
Y
Z

WEBSITE

EMAIL USED

USERNAME

PASSWORD

NOTES

WEBSITE

EMAIL USED

USERNAME

PASSWORD

NOTES

WEBSITE

EMAIL USED

USERNAME

PASSWORD

NOTES

A
B
C
D
E
F
G
H
I
J
K
L
M
N
O
P
Q
R
S
T
U
V
W
X
Y
Z

A
B
C | EMAIL USED
D | USERNAME
E
F | PASSWORD
G
H | NOTES
I
J | WEBSITE
K
L | EMAIL USED
M | USERNAME
N
O | PASSWORD
P
Q | NOTES
R
S | WEBSITE
T
U | EMAIL USED
V | USERNAME
W | PASSWORD
X
Y | NOTES
Z

WEBSITE

WEBSITE

EMAIL USED

USERNAME

PASSWORD

NOTES

WEBSITE

EMAIL USED

USERNAME

PASSWORD

NOTES

WEBSITE

EMAIL USED

USERNAME

PASSWORD

NOTES

A
B
C
D
E
F
G
H
I
J
K
L
M
N
O
P
Q
R
S
T
U
V
W
X
Y
Z

A
B
C
D
E
F
G
H
I
J
K
L
M
N
O
P
Q
R
S
T
U
V
W
X
Y
Z

WEBSITE

EMAIL USED

USERNAME

PASSWORD

NOTES

WEBSITE

EMAIL USED

USERNAME

PASSWORD

NOTES

WEBSITE

EMAIL USED

USERNAME

PASSWORD

NOTES

WEBSITE

EMAIL USED

USERNAME

PASSWORD

NOTES

WEBSITE

EMAIL USED

USERNAME

PASSWORD

NOTES

WEBSITE

EMAIL USED

USERNAME

PASSWORD

NOTES

A
B
C
D
E
F
G
H
I
J
K
L
M
N
O
P
Q
R
S
T
U
V
W
X
Y
Z

A
B
C
D
E
F
G
H
I
J
K
L
M
N
O
P
Q
R
S
T
U
V
W
X
Y
Z

WEBSITE

EMAIL USED

USERNAME

PASSWORD

NOTES

WEBSITE

EMAIL USED

USERNAME

PASSWORD

NOTES

WEBSITE

EMAIL USED

USERNAME

PASSWORD

NOTES

WEBSITE

EMAIL USED

USERNAME

PASSWORD

NOTES

WEBSITE

EMAIL USED

USERNAME

PASSWORD

NOTES

WEBSITE

EMAIL USED

USERNAME

PASSWORD

NOTES

A
B
C
D
E
F
G
H
I
J
K
L
M
N
O
P
Q
R
S
T
U
V
W
X
Y
Z

WEBSITE

EMAIL USED

USERNAME

PASSWORD

NOTES

WEBSITE

EMAIL USED

USERNAME

PASSWORD

NOTES

WEBSITE

EMAIL USED

USERNAME

PASSWORD

NOTES

WEBSITE

EMAIL USED

USERNAME

PASSWORD

NOTES

WEBSITE

EMAIL USED

USERNAME

PASSWORD

NOTES

WEBSITE

EMAIL USED

USERNAME

PASSWORD

NOTES

A
B
C
D
E
F
G
H
I
J
K
L
M
N
O
P
Q
R
S
T
U
V
W
X
Y
Z

A
B
WEBSITE
C
EMAIL USED

D
USERNAME

E
PASSWORD
F

G
NOTES
H

I
J
WEBSITE

K
EMAIL USED
L

M
USERNAME

N
PASSWORD
O

P
NOTES
Q

R
S
WEBSITE

T
EMAIL USED
U

V
USERNAME

W
PASSWORD
X

Y
NOTES
Z

WEBSITE

EMAIL USED

USERNAME

PASSWORD

NOTES

WEBSITE

EMAIL USED

USERNAME

PASSWORD

NOTES

WEBSITE

EMAIL USED

USERNAME

PASSWORD

NOTES

A
B
C
D
E
F
G
H
I
J
K
L
M
N
O
P
Q
R
S
T
U
V
W
X
Y
Z

WEBSITE

EMAIL USED

USERNAME

PASSWORD

NOTES

WEBSITE

EMAIL USED

USERNAME

PASSWORD

NOTES

WEBSITE

EMAIL USED

USERNAME

PASSWORD

NOTES

WEBSITE

EMAIL USED

USERNAME

PASSWORD

NOTES

WEBSITE

EMAIL USED

USERNAME

PASSWORD

NOTES

WEBSITE

EMAIL USED

USERNAME

PASSWORD

NOTES

A
B
C
D
E
F
G
H
I
J
K
L
M
N
O
P
Q
R
S
T
U
V
W
X
Y
Z

WEBSITE

EMAIL USED

USERNAME

PASSWORD

NOTES

WEBSITE

EMAIL USED

USERNAME

PASSWORD

NOTES

WEBSITE

EMAIL USED

USERNAME

PASSWORD

NOTES

WEBSITE

EMAIL USED

USERNAME

PASSWORD

NOTES

WEBSITE

EMAIL USED

USERNAME

PASSWORD

NOTES

WEBSITE

EMAIL USED

USERNAME

PASSWORD

NOTES

A
B
C
D
E
F
G
H
I
J
K
L
M
N
O
P
Q
R
S
T
U
V
W
X
Y
Z

A
B
C
D
E
F
G
H
I
J
K
L
M
N
O
P
Q
R
S
T
U
V
W
X
Y
Z

WEBSITE

EMAIL USED

USERNAME

PASSWORD

NOTES

WEBSITE

EMAIL USED

USERNAME

PASSWORD

NOTES

WEBSITE

EMAIL USED

USERNAME

PASSWORD

NOTES

WEBSITE

EMAIL USED

USERNAME

PASSWORD

NOTES

WEBSITE

EMAIL USED

USERNAME

PASSWORD

NOTES

WEBSITE

EMAIL USED

USERNAME

PASSWORD

NOTES

A
B
C
D
E
F
G
H
I
J
K
L
M
N
O
P
Q
R
S
T
U
V
W
X
Y
Z

A
B
WEBSITE
C EMAIL USED
D USERNAME
E
F PASSWORD
G
H NOTES
I
J
WEBSITE
K
L EMAIL USED
M USERNAME
N
O PASSWORD
P
Q NOTES
R
S
WEBSITE
T
U EMAIL USED
V USERNAME
W
X PASSWORD
Y NOTES
Z

WEBSITE

EMAIL USED

USERNAME

PASSWORD

NOTES

WEBSITE

EMAIL USED

USERNAME

PASSWORD

NOTES

WEBSITE

EMAIL USED

USERNAME

PASSWORD

NOTES

A
B
C
D
E
F
G
H
I
J
K
L
M
N
O
P
Q
R
S
T
U
V
W
X
Y
Z

A
B
C
D
E
F
G
H
I
J
K
L
M
N
O
P
Q
R
S
T
U
V
W
X
Y
Z

WEBSITE

EMAIL USED

USERNAME

PASSWORD

NOTES

WEBSITE

EMAIL USED

USERNAME

PASSWORD

NOTES

WEBSITE

EMAIL USED

USERNAME

PASSWORD

NOTES

WEBSITE

EMAIL USED

USERNAME

PASSWORD

NOTES

WEBSITE

EMAIL USED

USERNAME

PASSWORD

NOTES

WEBSITE

EMAIL USED

USERNAME

PASSWORD

NOTES

A
B
C
D
E
F
G
H
I
J
K
L
M
N
O
P
Q
R
S
T
U
V
W
X
Y
Z

A
B
C | **WEBSITE**

EMAIL USED

USERNAME

PASSWORD

NOTES

D
E
F
G
H
I
J
K
L
M
N
O
P
Q
R

WEBSITE

EMAIL USED

USERNAME

PASSWORD

NOTES

S
T
U
V
W
X
Y
Z

WEBSITE

EMAIL USED

USERNAME

PASSWORD

NOTES

WEBSITE

EMAIL USED

USERNAME

PASSWORD

NOTES

WEBSITE

EMAIL USED

USERNAME

PASSWORD

NOTES

WEBSITE

EMAIL USED

USERNAME

PASSWORD

NOTES

A
B
C
D
E
F
G
H
I
J
K
L
M
N
O
P
Q
R
S
T
U
V
W
X
Y
Z

A
B
C
D
E
F
G
H
I
J
K
L
M
N
O
P
Q
R
S
T
U
V
W
X
Y
Z

WEBSITE

EMAIL USED

USERNAME

PASSWORD

NOTES

WEBSITE

EMAIL USED

USERNAME

PASSWORD

NOTES

WEBSITE

EMAIL USED

USERNAME

PASSWORD

NOTES

WEBSITE

EMAIL USED

USERNAME

PASSWORD

NOTES

WEBSITE

EMAIL USED

USERNAME

PASSWORD

NOTES

WEBSITE

EMAIL USED

USERNAME

PASSWORD

NOTES

A
B
C
D
E
F
G
H
I
J
K
L
M
N
O
P
Q
R
S
T
U
V
W
X
Y
Z

A
B
C
D
E
F
G
H
I
J
K
L
M
N
O
P
Q
R
S
T
U
V
W
X
Y
Z

WEBSITE

EMAIL USED

USERNAME

PASSWORD

NOTES

WEBSITE

EMAIL USED

USERNAME

PASSWORD

NOTES

WEBSITE

EMAIL USED

USERNAME

PASSWORD

NOTES

WEBSITE

EMAIL USED

USERNAME

PASSWORD

NOTES

WEBSITE

EMAIL USED

USERNAME

PASSWORD

NOTES

WEBSITE

EMAIL USED

USERNAME

PASSWORD

NOTES

A
B
C
D
E
F
G
H
I
J
K
L
M
N
O
P
Q
R
S
T
U
V
W
X
Y
Z

A
B
WEBSITE

C
EMAIL USED

D
USERNAME

E
F
PASSWORD

G
H
NOTES

I
J
WEBSITE

K
L
EMAIL USED

M
USERNAME

N
O
PASSWORD

P
Q
NOTES

R
S
WEBSITE

T
U
EMAIL USED

V
USERNAME

W
X
PASSWORD

Y
Z
NOTES

WEBSITE

EMAIL USED

USERNAME

PASSWORD

NOTES

WEBSITE

EMAIL USED

USERNAME

PASSWORD

NOTES

WEBSITE

EMAIL USED

USERNAME

PASSWORD

NOTES

A
B
C
D
E
F
G
H
I
J
K
L
M
N
O
P
Q
R
S
T
U
V
W
X
Y
Z

A
B
C — EMAIL USED
D — USERNAME
E
F — PASSWORD
G
H — NOTES
I
J — WEBSITE
K
L — EMAIL USED
M — USERNAME
N
O — PASSWORD
P
Q — NOTES
R
S — WEBSITE
T
U — EMAIL USED
V — USERNAME
W — PASSWORD
X
Y — NOTES
Z

WEBSITE

EMAIL USED

USERNAME

PASSWORD

NOTES

WEBSITE

EMAIL USED

USERNAME

PASSWORD

NOTES

WEBSITE

EMAIL USED

USERNAME

PASSWORD

NOTES

WEBSITE

EMAIL USED

USERNAME

PASSWORD

NOTES

WEBSITE

EMAIL USED

USERNAME

PASSWORD

NOTES

WEBSITE

EMAIL USED

USERNAME

PASSWORD

NOTES

K

A
B
C
D
E
F
G
H
I
J
K
L
M
N
O
P
Q
R
S
T
U
V
W
X
Y
Z

WEBSITE

EMAIL USED

USERNAME

PASSWORD

NOTES

WEBSITE

EMAIL USED

USERNAME

PASSWORD

NOTES

WEBSITE

EMAIL USED

USERNAME

PASSWORD

NOTES

WEBSITE

EMAIL USED

USERNAME

PASSWORD

NOTES

WEBSITE

EMAIL USED

USERNAME

PASSWORD

NOTES

WEBSITE

EMAIL USED

USERNAME

PASSWORD

NOTES

A
B
C
D
E
F
G
H
I
J
K
L
M
N
O
P
Q
R
S
T
U
V
W
X
Y
Z

A
B
WEBSITE
C EMAIL USED
D USERNAME
E
F PASSWORD
G
H NOTES
I
J **WEBSITE**
K
L EMAIL USED
M USERNAME
N
O PASSWORD
P
Q NOTES
R
S **WEBSITE**
T
U EMAIL USED
V USERNAME
W
X PASSWORD
Y NOTES
Z

WEBSITE

EMAIL USED

USERNAME

PASSWORD

NOTES

WEBSITE

EMAIL USED

USERNAME

PASSWORD

NOTES

WEBSITE

EMAIL USED

USERNAME

PASSWORD

NOTES

A
B
C
D
E
F
G
H
I
J
K
L
M
N
O
P
Q
R
S
T
U
V
W
X
Y
Z

WEBSITE

EMAIL USED

USERNAME

PASSWORD

NOTES

WEBSITE

EMAIL USED

USERNAME

PASSWORD

NOTES

WEBSITE

EMAIL USED

USERNAME

PASSWORD

NOTES

WEBSITE

EMAIL USED

USERNAME

PASSWORD

NOTES

WEBSITE

EMAIL USED

USERNAME

PASSWORD

NOTES

WEBSITE

EMAIL USED

USERNAME

PASSWORD

NOTES

A
B
C
D
E
F
G
H
I
J
K
L
M
N
O
P
Q
R
S
T
U
V
W
X
Y
Z

A
B
WEBSITE
C
EMAIL USED
D
USERNAME
F
PASSWORD
F
G
NOTES
H
I
J
WEBSITE
K
EMAIL USED
L
M
USERNAME
N
PASSWORD
O
P
NOTES
Q
R
S
WEBSITE
T
EMAIL USED
U
V
USERNAME
W
PASSWORD
X
Y
NOTES
Z

WEBSITE

EMAIL USED

USERNAME

PASSWORD

NOTES

WEBSITE

EMAIL USED

USERNAME

PASSWORD

NOTES

WEBSITE

EMAIL USED

USERNAME

PASSWORD

NOTES

A
B
C
D
E
F
G
H
I
J
K
L
M
N
O
P
Q
R
S
T
U
V
W
X
Y
Z

A	**WEBSITE**
B	
C	EMAIL USED
D	USERNAME
E	
F	PASSWORD
G	
H	NOTES
I	
J	**WEBSITE**
K	
L	EMAIL USED
M	USERNAME
N	
O	PASSWORD
P	
Q	NOTES
R	
S	**WEBSITE**
T	
U	EMAIL USED
V	USERNAME
W	
X	PASSWORD
Y	NOTES
Z	

WEBSITE

EMAIL USED

USERNAME

PASSWORD

NOTES

WEBSITE

EMAIL USED

USERNAME

PASSWORD

NOTES

WEBSITE

EMAIL USED

USERNAME

PASSWORD

NOTES

A
B
C
D
E
F
G
H
I
J
K
L
M
N
O
P
Q
R
S
T
U
V
W
X
Y
Z

A
B
WEBSITE

C
EMAIL USED

D
USERNAME

E
F
PASSWORD

G
H
NOTES

I
J
WEBSITE

K
L
EMAIL USED

M
USERNAME

N
PASSWORD

O
P
NOTES

Q
R
S
WEBSITE

T
U
EMAIL USED

V
USERNAME

W
PASSWORD

X
Y
NOTES

Z

WEBSITE

EMAIL USED

USERNAME

PASSWORD

NOTES

WEBSITE

EMAIL USED

USERNAME

PASSWORD

NOTES

WEBSITE

EMAIL USED

USERNAME

PASSWORD

NOTES

A
B
C
D
E
F
G
H
I
J
K
L
M
N
O
P
Q
R
S
T
U
V
W
X
Y
Z

A
B
C
D
E
F
G
H
I
J
K
L
M
N
O
P
Q
R
S
T
U
V
W
X
Y
Z

WEBSITE

EMAIL USED

USERNAME

PASSWORD

NOTES

WEBSITE

EMAIL USED

USERNAME

PASSWORD

NOTES

WEBSITE

EMAIL USED

USERNAME

PASSWORD

NOTES

WEBSITE

EMAIL USED

USERNAME

PASSWORD

NOTES

WEBSITE

EMAIL USED

USERNAME

PASSWORD

NOTES

WEBSITE

EMAIL USED

USERNAME

PASSWORD

NOTES

A
B
WEBSITE

C
EMAIL USED

D
USERNAME

E
PASSWORD
F

G
NOTES
H

I
J
WEBSITE

K
EMAIL USED
L

M
USERNAME

N
PASSWORD
O

P
NOTES
Q

R
S
WEBSITE

T
EMAIL USED
U

V
USERNAME

W
PASSWORD
X

Y
NOTES
Z

WEBSITE

EMAIL USED

USERNAME

PASSWORD

NOTES

WEBSITE

EMAIL USED

USERNAME

PASSWORD

NOTES

WEBSITE

EMAIL USED

USERNAME

PASSWORD

NOTES

A
B
C
D
E
F
G
H
I
J
K
L
M
N
O
P
Q
R
S
T
U
V
W
X
Y
Z

A
B
WEBSITE
C
EMAIL USED

D
USERNAME

E
PASSWORD
F

G
NOTES
H

I
J
WEBSITE
K
EMAIL USED
L

M
USERNAME

N
PASSWORD
O

P
NOTES
Q

R
S
WEBSITE
T
EMAIL USED
U

V
USERNAME

W
PASSWORD
X

Y
NOTES
Z

WEBSITE

EMAIL USED

USERNAME

PASSWORD

NOTES

WEBSITE

EMAIL USED

USERNAME

PASSWORD

NOTES

WEBSITE

EMAIL USED

USERNAME

PASSWORD

NOTES

A
B
C
D
E
F
G
H
I
J
K
L
M
N
O
P
Q
R
S
T
U
V
W
X
Y
Z

A
B
C
D
E
F
G
H
I
J
K
L
M
N
O
P
Q
R
S
T
U
V
W
X
Y
Z

WEBSITE

EMAIL USED

USERNAME

PASSWORD

NOTES

WEBSITE

EMAIL USED

USERNAME

PASSWORD

NOTES

WEBSITE

EMAIL USED

USERNAME

PASSWORD

NOTES

WEBSITE

EMAIL USED

USERNAME

PASSWORD

NOTES

WEBSITE

EMAIL USED

USERNAME

PASSWORD

NOTES

WEBSITE

EMAIL USED

USERNAME

PASSWORD

NOTES

A
B
C   EMAIL USED
D   USERNAME
E
F   PASSWORD
G
H   NOTES
I
J   WEBSITE
K
L   EMAIL USED
M   USERNAME
N
O   PASSWORD
P   NOTES
Q
R
S   WEBSITE
T
U   EMAIL USED
V   USERNAME
W   PASSWORD
X
Y   NOTES
Z

WEBSITE

EMAIL USED

USERNAME

PASSWORD

NOTES

WEBSITE

EMAIL USED

USERNAME

PASSWORD

NOTES

WEBSITE

EMAIL USED

USERNAME

PASSWORD

NOTES

A
B
C
D
E
F
G
H
I
J
K
L
M
N
O
P
Q
R
S
T
U
V
W
X
Y
Z

WEBSITE

EMAIL USED

USERNAME

PASSWORD

NOTES

WEBSITE

EMAIL USED

USERNAME

PASSWORD

NOTES

WEBSITE

EMAIL USED

USERNAME

PASSWORD

NOTES

WEBSITE

EMAIL USED

USERNAME

PASSWORD

NOTES

WEBSITE

EMAIL USED

USERNAME

PASSWORD

NOTES

WEBSITE

EMAIL USED

USERNAME

PASSWORD

NOTES

A
B
WEBSITE
C
EMAIL USED
D
USERNAME
E
PASSWORD
F
G
NOTES
H
I
J
WEBSITE
K
L
EMAIL USED
M
USERNAME
N
PASSWORD
O
P
NOTES
Q
R
S
WEBSITE
T
U
EMAIL USED
V
USERNAME
W
PASSWORD
X
Y
NOTES
Z

WEBSITE

EMAIL USED

USERNAME

PASSWORD

NOTES

WEBSITE

EMAIL USED

USERNAME

PASSWORD

NOTES

WEBSITE

EMAIL USED

USERNAME

PASSWORD

NOTES

A B C D E F G H I J K L M N O P **Q** R S T U V W X Y Z

A
B
C
D
E
F
G
H
I
J
K
L
M
N
O
P
Q
R
S
T
U
V
W
X
Y
Z

WEBSITE

EMAIL USED

USERNAME

PASSWORD

NOTES

WEBSITE

EMAIL USED

USERNAME

PASSWORD

NOTES

WEBSITE

EMAIL USED

USERNAME

PASSWORD

NOTES

WEBSITE

EMAIL USED

USERNAME

PASSWORD

NOTES

WEBSITE

EMAIL USED

USERNAME

PASSWORD

NOTES

WEBSITE

EMAIL USED

USERNAME

PASSWORD

NOTES

A
B
C | **WEBSITE**
D |
E | EMAIL USED
F |
G | USERNAME
H |
I | PASSWORD
J |
K | NOTES

A
B
C
D
E
F
G
H
I
J
K
L
M
N
O
P
Q
R
S
T
U
V
W
X
Y
Z

WEBSITE

EMAIL USED

USERNAME

PASSWORD

NOTES

WEBSITE

EMAIL USED

USERNAME

PASSWORD

NOTES

WEBSITE

EMAIL USED

USERNAME

PASSWORD

NOTES

WEBSITE

EMAIL USED

USERNAME

PASSWORD

NOTES

WEBSITE

EMAIL USED

USERNAME

PASSWORD

NOTES

WEBSITE

EMAIL USED

USERNAME

PASSWORD

NOTES

A
B
C
D
E
F
G
H
I
J
K
L
M
N
O
P
Q
R
S
T
U
V
W
X
Y
Z

A
B
WEBSITE

C
EMAIL USED

D
USERNAME

E
F
PASSWORD

G
H
NOTES

I
J
WEBSITE

K
L
EMAIL USED

M
USERNAME

N
O
PASSWORD

P
Q
NOTES

R
S
WEBSITE

T
U
EMAIL USED

V
USERNAME

W
X
PASSWORD

Y
Z
NOTES

WEBSITE

EMAIL USED

USERNAME

PASSWORD

NOTES

WEBSITE

EMAIL USED

USERNAME

PASSWORD

NOTES

WEBSITE

EMAIL USED

USERNAME

PASSWORD

NOTES

A
B
C
D
E
F
G
H
I
J
K
L
M
N
O
P
Q
R
S
T
U
V
W
X
Y
Z

A
B
C | EMAIL USED |
D | USERNAME |
E
F | PASSWORD |
G
H | NOTES |
I
J | WEBSITE |
K
L | EMAIL USED |
M | USERNAME |
N
O | PASSWORD |
P
Q | NOTES |
R
S | WEBSITE |
T
U | EMAIL USED |
V | USERNAME |
W
X | PASSWORD |
Y | NOTES |
Z

WEBSITE

EMAIL USED

USERNAME

PASSWORD

NOTES

WEBSITE

EMAIL USED

USERNAME

PASSWORD

NOTES

WEBSITE

EMAIL USED

USERNAME

PASSWORD

NOTES

A
B
C
D
E
F
G
H
I
J
K
L
M
N
O
P
Q
R
S
T
U
V
W
X
Y
Z

A
B
WEBSITE
C   EMAIL USED
D   USERNAME
E
F   PASSWORD
G
H   NOTES
I
J   ## WEBSITE
K
L   EMAIL USED
M   USERNAME
N   PASSWORD
O
P   NOTES
Q
R
S   ## WEBSITE
T
U   EMAIL USED
V   USERNAME
W   PASSWORD
X
Y   NOTES
Z

WEBSITE

EMAIL USED

USERNAME

PASSWORD

NOTES

WEBSITE

EMAIL USED

USERNAME

PASSWORD

NOTES

WEBSITE

EMAIL USED

USERNAME

PASSWORD

NOTES

A
B
C
D
E
F
G
H
I
J
K
L
M
N
O
P
Q
R
S
T
U
V
W
X
Y
Z

A
B
C | WEBSITE

D | EMAIL USED

E | USERNAME
F
G | PASSWORD
H
I | NOTES
J
K | WEBSITE
L | EMAIL USED
M | USERNAME
N | PASSWORD
O
P | NOTES
Q
R
S | WEBSITE
T
U | EMAIL USED
V | USERNAME
W | PASSWORD
X
Y | NOTES
Z

WEBSITE

EMAIL USED

USERNAME

PASSWORD

NOTES

WEBSITE

EMAIL USED

USERNAME

PASSWORD

NOTES

WEBSITE

EMAIL USED

USERNAME

PASSWORD

NOTES

A
B
C
D
E
F
G
H
I
J
K
L
M
N
O
P
Q
R
S
T
U
V
W
X
Y
Z

A
B
C — EMAIL USED
D — USERNAME
E
F — PASSWORD
G
H — NOTES
I
J — WEBSITE
K
L — EMAIL USED
M — USERNAME
N
O — PASSWORD
P
Q — NOTES
R
S — WEBSITE
T
U — EMAIL USED
V — USERNAME
W
X — PASSWORD
Y — NOTES
Z

WEBSITE

EMAIL USED

USERNAME

PASSWORD

NOTES

WEBSITE

EMAIL USED

USERNAME

PASSWORD

NOTES

WEBSITE

EMAIL USED

USERNAME

PASSWORD

NOTES

WEBSITE

EMAIL USED

USERNAME

PASSWORD

NOTES

WEBSITE

EMAIL USED

USERNAME

PASSWORD

NOTES

WEBSITE

EMAIL USED

USERNAME

PASSWORD

NOTES

A
B
C
D
E
F
G
H
I
J
K
L
M
N
O
P
Q
R
S
T
U
V
W
X
Y
Z

A
B
C   EMAIL USED
D   USERNAME
E
F   PASSWORD

G
H   NOTES

I
J   WEBSITE
K
L   EMAIL USED
M   USERNAME
N
O   PASSWORD

P
Q   NOTES

R
S   WEBSITE
T
U   EMAIL USED
V   USERNAME
W   PASSWORD
X
Y   NOTES
Z

WEBSITE

EMAIL USED

USERNAME

PASSWORD

NOTES

WEBSITE

EMAIL USED

USERNAME

PASSWORD

NOTES

WEBSITE

EMAIL USED

USERNAME

PASSWORD

NOTES

A
B
C | WEBSITE
D
E | EMAIL USED
F
G | USERNAME
H
I | PASSWORD
J
K | NOTES
L
M
N | WEBSITE
O
P | EMAIL USED
Q
R | USERNAME
S
T | PASSWORD
U
V | NOTES
W
X
Y | WEBSITE
Z

- A
- B
- C — WEBSITE
- D
- E — EMAIL USED
- F
- G — USERNAME
- H
- I — PASSWORD
- J
- K — NOTES

- WEBSITE
- EMAIL USED
- USERNAME
- PASSWORD
- NOTES

- WEBSITE
- EMAIL USED
- USERNAME
- PASSWORD
- NOTES

WEBSITE

EMAIL USED

USERNAME

PASSWORD

NOTES

WEBSITE

EMAIL USED

USERNAME

PASSWORD

NOTES

WEBSITE

EMAIL USED

USERNAME

PASSWORD

NOTES

A
B
C
D
E
F
G
H
I
J
K
L
M
N
O
P
Q
R
S
T
U
V
W
X
Y
Z

A
B
C   **WEBSITE**
D
E   EMAIL USED
F
G   USERNAME
H
I   PASSWORD
J
K   NOTES
L
M

WEBSITE

EMAIL USED

USERNAME

PASSWORD

NOTES

WEBSITE

EMAIL USED

USERNAME

PASSWORD

NOTES

WEBSITE

EMAIL USED

USERNAME

PASSWORD

NOTES

WEBSITE

EMAIL USED

USERNAME

PASSWORD

NOTES

WEBSITE

EMAIL USED

USERNAME

PASSWORD

NOTES

A
B
C
D
E
F
G
H
I
J
K
L
M
N
O
P
Q
R
S
T
U
V
W
X
Y
Z

A
B
C
D
E
F
G
H
I
J
K
L
M
N
O
P
Q
R
S
T
U
V
W
X
Y
Z

WEBSITE

EMAIL USED

USERNAME

PASSWORD

NOTES

WEBSITE

EMAIL USED

USERNAME

PASSWORD

NOTES

WEBSITE

EMAIL USED

USERNAME

PASSWORD

NOTES

| WEBSITE |

| EMAIL USED |
| USERNAME |
| PASSWORD |
| NOTES |

| WEBSITE |

| EMAIL USED |
| USERNAME |
| PASSWORD |
| NOTES |

| WEBSITE |

| EMAIL USED |
| USERNAME |
| PASSWORD |
| NOTES |

A B C D E F G H I J K L M N O P Q R S T U V **W** X Y Z

A
B
WEBSITE

C EMAIL USED

D USERNAME

E
F PASSWORD

G
H NOTES

I
J
WEBSITE
K

L EMAIL USED

M USERNAME

N
O PASSWORD

P
Q NOTES

R
S
WEBSITE
T

U EMAIL USED

V USERNAME

W
X PASSWORD

Y NOTES
Z

WEBSITE

EMAIL USED

USERNAME

PASSWORD

NOTES

WEBSITE

EMAIL USED

USERNAME

PASSWORD

NOTES

WEBSITE

EMAIL USED

USERNAME

PASSWORD

NOTES

A
B
C
D
E
F
G
H
I
J
K
L
M
N
O
P
Q
R
S
T
U
V
W
X
Y
Z

A
B
C
D
E
F
G
H
I
J
K
L
M
N
O
P
Q
R
S
T
U
V
W
X
Y
Z

WEBSITE

EMAIL USED

USERNAME

PASSWORD

NOTES

WEBSITE

EMAIL USED

USERNAME

PASSWORD

NOTES

WEBSITE

EMAIL USED

USERNAME

PASSWORD

NOTES

WEBSITE

EMAIL USED

USERNAME

PASSWORD

NOTES

WEBSITE

EMAIL USED

USERNAME

PASSWORD

NOTES

WEBSITE

EMAIL USED

USERNAME

PASSWORD

NOTES

A
B
C   WEBSITE

D   EMAIL USED

E   USERNAME

F   PASSWORD

G
H   NOTES

I
J   WEBSITE
K
L   EMAIL USED

M   USERNAME

N
O   PASSWORD

P
Q   NOTES

R
S   WEBSITE
T
U   EMAIL USED

V   USERNAME

W   PASSWORD
X
Y   NOTES
Z

WEBSITE

EMAIL USED

USERNAME

PASSWORD

NOTES

WEBSITE

EMAIL USED

USERNAME

PASSWORD

NOTES

WEBSITE

EMAIL USED

USERNAME

PASSWORD

NOTES

A
B
| WEBSITE |

C EMAIL USED

D USERNAME

E
F PASSWORD

G
H NOTES

I
J
K
| WEBSITE |

L EMAIL USED

M USERNAME

N
O PASSWORD

P
Q NOTES

R
S
T
| WEBSITE |

U EMAIL USED

V USERNAME

W
X PASSWORD

Y
Z NOTES

WEBSITE

EMAIL USED

USERNAME

PASSWORD

NOTES

WEBSITE

EMAIL USED

USERNAME

PASSWORD

NOTES

WEBSITE

EMAIL USED

USERNAME

PASSWORD

NOTES

A
B
C
D
E
F
G
H
I
J
K
L
M
N
O
P
Q
R
S
T
U
V
W
X
Y
Z

A
B
C | WEBSITE |
D
| EMAIL USED |
E
| USERNAME |
F
| PASSWORD |
G
H | NOTES |
I
J
K | WEBSITE |
L
| EMAIL USED |
M
| USERNAME |
N
O | PASSWORD |
P
Q | NOTES |
R
S
T | WEBSITE |
U
| EMAIL USED |
V
| USERNAME |
W
X | PASSWORD |
Y
Z | NOTES |

WEBSITE

EMAIL USED

USERNAME

PASSWORD

NOTES

WEBSITE

EMAIL USED

USERNAME

PASSWORD

NOTES

WEBSITE

EMAIL USED

USERNAME

PASSWORD

NOTES

A
B
WEBSITE
C EMAIL USED
D USERNAME
E
F PASSWORD
G
H NOTES
I
J
WEBSITE
K
L EMAIL USED
M USERNAME
N
O PASSWORD
P
Q NOTES
R
S
WEBSITE
T
U EMAIL USED
V USERNAME
W
X PASSWORD
Y NOTES
Z

WEBSITE

EMAIL USED

USERNAME

PASSWORD

NOTES

WEBSITE

EMAIL USED

USERNAME

PASSWORD

NOTES

WEBSITE

EMAIL USED

USERNAME

PASSWORD

NOTES

A
B
C | EMAIL USED
D | USERNAME
E
F | PASSWORD
G
H | NOTES
I

WEBSITE

J
K
L | EMAIL USED
M | USERNAME
N
O | PASSWORD
P
Q | NOTES
R

WEBSITE

S
T
U | EMAIL USED
V | USERNAME
W
X | PASSWORD
Y
Z | NOTES

WEBSITE (top)

| WEBSITE |

| EMAIL USED |
| USERNAME |
| PASSWORD |
| NOTES |

| WEBSITE |

| EMAIL USED |
| USERNAME |
| PASSWORD |
| NOTES |

| WEBSITE |

| EMAIL USED |
| USERNAME |
| PASSWORD |
| NOTES |

A
B
C
D
E
F
G
H
I
J
K
L
M
N
O
P
Q
R
S
T
U
V
W
X
Y
Z

NOTES

NOTES

NOTES

NOTES

Copyright ©
All rights reserved. No part of this publication may be reproduced, distributed,
or transmitted in any form or by any means, including photocopying, recording,
or other electronic or mechanical methods, without the prior written permission
of the publisher, except in the case of brief quotations embodied in critical reviews
and certain other noncommercial uses permitted by copyright law.

Made in the USA
Columbia, SC
06 June 2024